# HARFLARNING HIKOYASI

## THE NUMBER STORY

---

### SMALL BOOK ONE

ENGLISH - UZBEK

---

*Numbers Teach Children*
*Their Number Names*

written and illustrated by

# MISS ANNA

Early Reader Edition of *The Number Story 1*
Bronze Medal Winner, 2016 Wishing Shelf Book Award

Library of Congress Control Number: 2018902040

Names: Miss Anna, author.
Title: Number story : numbers teach children their number names / Miss Anna.
Description: Portland, OR: Lumpy Publishing, 2018.
Identifiers: ISBN 978-1-949320-01-5| LCCN 2018902040
Summary: The pictures and rhymes present stories which introduce numbers 0-10.
Subjects: LCSH Numeration—English—Uzbek--Pictorial works--Juvenile literature. | BISAC JUVENILE NONFICTION /
Languages: English—Uzbek
Classification: LCC QA141.3 .M57 2018 | DDC 513—dc23

Publisher: Lumpy Publishing
Website: www.missannabooks.com
Email: missanna@missannabooks.com

Paperback: ISBN 978-1-949320-01-5
Printed in the U.S.A.    1 3 5 7 9 10 8 6 4 2

Raqamlarning nomini o'rganishni xohlaysizmi?

It is very easy and a lot of fun!

Bu judaham qiziq va qulay!

Say-along our little jingle

Birgalikda kuylaymiz kichik hikoyamizni!

starting from Number One!

**Birinchi raqamdan boshlaymiz!**

# 1

ONE  looks like my one finger.

## BIR

Mening bitta barmoq'imga

o'xshaydi.

ONE!
BIR!

2

TWO     trails a tail.

IKKI

dumini sudraydi.

A TAIL! DUMINI!

# 3

THREE has bumps.

UCH

do'ng joylari bor.

BUMPY! PAST BALAND!

4

FOUR  carries a sail.

TO'RT

yelkan olib yuradi.

4
A SAIL!
YELKAN!

# 5

FIVE   is a racing track.

BESH

bu poyga yo'li.

VROOM
VROOM!
1

# 6

SIX   curves like a snail.

OLTI

shillik qurt kabi egiladi.

A SNAIL!   SHILLIQ QURT!

7

SEVEN has a sharp angle.

YETTI

o'tkir burchagi bor.

BE CAREFUL! IT'S SHARP!

Judaham Ehtiyot bo'ling! bu O'tkir!

8

EIGHT   is rollercoaster rails.

SAKKIZ

bu rollercoaster relslari.kabi.

URAAA!
YIPPEE!

NINE   is a bubble on a stick.

TO'QQIZ

bu Tayoq ustidagi balon.

A BUBBLE!

BALON!

# 10

TEN  is an eye of a whale.

O'N

bu kitning bir ko'zidir.

HELLO!

SALOM!

And
Va

0

ZERO   is an empty pail.

NOL

bu bo'sh paqir.

IT'S
EMPTY!
BU BO'SH!

Thank you for playing with us today.

We had a lot of fun too!

Bugun biz bilan o'ynaganingiz
uchun rahmat.

Bizham xursandchilik qildik!

We are your Number friends,
Zero to Ten,
Who will be here for you~
Bizlar sizning Raqamli do'stlaringizmiz
Noldan o'ngacha.
Biz har doim hozirmiz sizlar uchun!

Bye-bye now!
See you again soon!
Hozircha xayr!
Yana ko'rishguncha!

The Numbers are *SINGING* too!

To sing-a-long, look for Miss Anna Number Story
at your favorite music store like iTUNES.

MP3

Numbers 0-10
IDENTIFYING
& COUNTING

Numbers 11-20
& Ordinals

first, second, third...

Numbers 0-100
& Place Values

ones, tens, hundreds...

About Clocks
& Telling Time

hours, minutes, seconds...

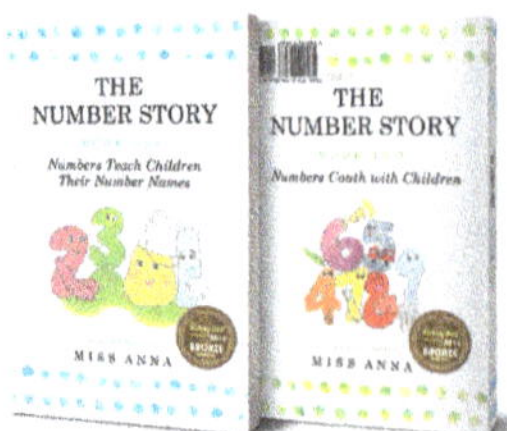

Number Story 1 & 2

isbn: 978-0-996216-48-7

Number Story 3 & 4

isbn: 978-1-945977-01-5

Number Story 5 & 6

isbn: 978-1-945977-06-0

Number Story 7 & 8

isbn: 978-1-949320-40-4

For more Miss Anna books to love,
visit us at

w w w . m i s s a n n a b o o k s . c o m

Numbers are working hard all over the world!
*Come Travel the World with Us!*